Mary Cleave

First woman astronaut to fly into space dies at 76

By

Bert Rivers

Table of contents

Biography and Early Life:

Mary Cleave, a groundbreaking astronaut, was born in Southampton, New York, in 1947 to Howard and Barbara Cleave, both educators. She embarked on her academic journey by earning a Bachelor of Science degree in biological sciences from Colorado State University in 1969. Cleave's educational pursuits continued with a Master of Science in microbial ecology and a doctorate in civil and environmental engineering from Utah State University in 1975 and 1979, respectively.

In May 1980, Cleave achieved a significant milestone by being selected as an astronaut.

Her space journey took off in 1985 with the STS-61B mission aboard the space shuttle Atlantis, marking her first venture into space. Tragedy struck with the Challenger explosion in 1986, but Cleave went on to become the first woman to fly in space after the incident, undertaking the STS-30 mission in 1989.

Beyond space, Cleave's contributions extended to NASA's Goddard Space Flight Center in Maryland, where she delved into monitoring global ocean vegetation. She later assumed roles such as Deputy Associate Administrator for Advanced Planning and Associate Administrator for NASA's Science Mission Directorate, leaving an indelible mark on Earth and space exploration.

Academic Prowess and NASA Odyssey

Academic Career:

Mary Cleave's academic journey unfolded from September 1971 to June 1980, encompassing roles as a graduate researcher, research phycologist, and research engineer at Utah State University's Ecology Center and Utah Water Research Laboratory. Her research spanned diverse ecological domains, including algal productivity in cold desert soil crusts, algal removal techniques, and the impact of salinity and oil shale leachates on freshwater phytoplankton. Notably, she contributed to the Surface Impoundment Assessment document and FORTRAN computer program for data processing. Cleave's commitment extended to establishing an algal bioassay center and facilitating bioassay techniques workshops for the Intermountain West.

NASA Career:

Transitioning to NASA, Cleave, selected as an astronaut in May 1980, showcased her technical acumen in various capacities. From flight software verification in the Shuttle Avionics Integration Laboratory to serving as CAPCOM on five Space Shuttle flights, her contributions were integral. A veteran of two space flights—STS-61-B (November 26 to December 3, 1985) and STS-30 (May 4–8, 1989)—Cleave logged over 10 days in space, orbiting the Earth 172 times and covering 3.94 million miles.

Cleave in 2006:

In May 1991, Cleave transitioned to NASA's Goddard Space Flight Center, where she took on the role of Project Manager for SeaWiFS, an ocean color sensor monitoring global vegetation. Her leadership continued as the Associate Administrator for NASA's Science Mission Directorate at NASA Headquarters in Washington, D.C. Cleave's tenure showcased her dedication to advancing scientific exploration.

Spaceflight Experience:

STS-61-B, launched from Kennedy Space Center, Florida, witnessed Cleave and her crew deploying communications satellites, conducting spacewalks, and operating various experiments. On STS-30, the crew successfully deployed the Magellan Venus-exploration spacecraft, a landmark mission mapping over 95% of Venus's surface. These missions solidified Cleave's legacy in space exploration, contributing valuable insights to planetary science.

Mary Cleave's journey weaves a narrative of academic brilliance seamlessly merging with a stellar NASA career, leaving an indelible mark on the realms of ecology and space exploration.

Journey Beyond the Stars

In May 1980, Mary Cleave's odyssey into the cosmos began when she was selected as an astronaut. Her technical prowess shone through various roles, from verifying flight software in the Shuttle Avionics Integration Laboratory to being a spacecraft communicator on five space shuttle flights. Cleave's expertise extended to crafting malfunction procedures and designing crew equipment.

Launching into the celestial realm on STS-61B aboard the Atlantis on Nov. 26, 1985, Cleave and her crew orchestrated a symphony of space endeavors. From deploying communications satellites to executing spacewalks showcasing station construction techniques, they navigated the cosmos with precision. The mission included operating experiments for McDonnell Douglas and Telesat, along with testing the Orbiter Experiments Digital Autopilot.

STS-30 marked Cleave's second mission on Atlantis, launching on May 4, 1989. This four-day expedition witnessed the successful deployment of the Magellan Venus exploration spacecraft, a pioneering moment as the first planetary probe dispatched from a space shuttle. Mapping over 95% of Venus's surface, the crew also engaged in secondary experiments spanning indium crystal growth, electrical storm observations, and Earth studies.

Transitioning from NASA's Johnson Space Center to the Goddard Space Flight Center in May 1991, Cleave delved into global environmental monitoring as the project manager for SeaWiFS. This ocean color sensor tracked vegetation worldwide, showcasing her commitment to Earth-centric scientific pursuits.

March 2000 saw Cleave assuming the role of deputy associate administrator for advanced planning at NASA's Headquarters in Washington. From August 2005 to February 2007, she ascended to the position of associate administrator for NASA's Science Mission Directorate, steering diverse research programs across Earth, space weather, the solar system, and the universe.

Cleave's illustrious career garnered accolades, including two NASA Space Flight medals, two NASA Exceptional Service medals, and an American Astronautical Society Flight Achievement Award. Her contributions were further recognized with a NASA Exceptional Achievement Medal and the prestigious title of NASA Engineer of the Year.

In February 2007, Mary Cleave bid adieu to NASA, leaving behind a legacy etched in the stars, symbolizing a remarkable journey of exploration and scientific leadership.

Union with James Crockett:

On May 11, 1829, in Jefferson, Indiana, Mary Leah embarked on a new chapter of her life as she exchanged vows with James Crockett. This union marked a significant moment, weaving her story into the fabric of

Trivia

Siblings and Family Roots:

- Mary Cleave's familial bonds included an older sister, Trudy, and a younger sister, Barbara, enriching her journey with shared memories and connections.

Educational Milestones:

- In 1969, Mary earned her Bachelor of Science degree in Biological Sciences from Colorado State University, paving the way for a scholarly path. Subsequently, she delved into advanced studies, achieving a Master of Science in Microbial Ecology in

1975 and attaining a Doctorate in Civil and Environmental Engineering in 1979, both from Utah State University.

High School Graduation:

- Mary's academic journey began at Great Neck North High School in Great Neck, New York, where she graduated in 1965, laying the foundation for her future endeavors.

Commemorative Stamp Recognition:

- The year 1995 marked a unique honor for Mary Cleave as she graced a postal stamp in Azerbaijan's series commemorating the 25th anniversary of the first manned moon landing, becoming a symbol of space exploration.

Heyden Distinguished Lecture Series:

- On January 18, 2009, Mary Cleave took the stage as the inaugural speaker in the Heyden Distinguished Lecture Series at Georgetown University. She shared insights into her education and career, offering a glimpse into her 1985 shuttle mission through an original film.

Affiliations and Memberships:

- Mary Cleave was an active member of various esteemed organizations, including the Association of Space Explorers, Sigma Xi, the Society for Professional Engineers, Tau Beta Pi, Women in Aerospace, and Tri Beta, Beta Beta Beta. These affiliations underscore her commitment to professional excellence and collaborative exploration.

Personal Portrait of Mary Cleave

Marital Status:

Unveiling the personal facet of Mary Cleave, her marital status stands as unmarried, allowing her journey to unfold with a focus on professional pursuits and personal passions.

Children:

Her life's narrative includes a blank page in the chapter of parenthood, as Mary Cleave did not have children, choosing to channel her energies into a diverse array of interests and accomplishments.

Recreational Pursuits:

Beyond the cosmic expanse, Mary finds joy in earthly adventures. Cross-country and downhill skiing, sailing, hiking, and camping constitute the palette of her recreational canvas, illustrating a well-rounded life embracing nature's wonders.

Organizational Ties:

Mary Cleave's professional tapestry is woven with affiliations, including the Texas Society of Professional Engineers, the Water Pollution Control Federation, Tri-Beta, Sigma Xi, and Tau Beta Pi. Additionally, she holds the role of an associate member in the esteemed American Society of Civil Engineers.

Special Honors:

The accolades adorning Mary's journey reflect her stellar contributions—recipient of the prestigious NASA Space Flight

Medal in 1985 and the NASA Exceptional Service Medal in 1988. These honors encapsulate the recognition of her exceptional endeavors in the realm of space exploration.

Starry Struggles and Orbital Wisdom:

In the cosmic ballet of space, Mary Cleave's defining moment wasn't the breathtaking 17,000 mph ascent into the cosmos but a fleeting struggle to locate her navigational stars on the Atlantis Space Shuttle's orbit on November 26, 1985. Above Earth's atmosphere, amidst the absence of light pollution, Mary faced a disorienting vastness. Yet, with resilience, she found her celestial markers, a crucial task for a flight engineer anticipating potential navigational challenges.

Earthly Foundations of a Spacefarer:

Mary's earthly pursuits were as daring as her cosmic ventures. From playing with model airplanes over dolls to earning her pilot's license at 17 before her driver's license, she soared into a world of fast-paced, supersonic jet trainers. Her aviation passion fueled her journey into NASA's astronaut corps in 1980, becoming part of only the second class of shuttle astronauts.

Mary's legacy extends beyond Earth, with two shuttle missions—STS-61B (1985) and STS-30 (1989). Her diverse research, from measuring plant life from space to studying carbon uptake in western deserts, showcased a scientific finesse matched by her hands-on problem-solving, even fixing a broken shuttle toilet.

Continued Exploration and Advocacy:

Retired from NASA in 2007, Mary, at 72, continues her journey. Engaged in research projects like the OPAL project with USU's Space Dynamics Laboratory, she also dedicates time to community service, acknowledging the pivotal role of the League of Women Voters in championing Title IX.

Trailblazing Legacy and Unwavering Wonder:

Mary Cleave's story resonates with a sense of wonder and an unwavering spirit of exploration. Whether navigating celestial realms or organizing candidate debates on Earth, she epitomizes a pioneer forging paths for others, emphasizing that progress, both celestial and terrestrial, demands perseverance through ups and downs.

As Mary aptly puts it, "I don't differentiate between the two. As I see it, you got to understand your neighborhood, or you're going to get into trouble." In her quest to "follow the water," Mary Cleave continues inspiring, exploring, and breaking barriers.

Mary Cleave, passed away at 76

Renowned astronaut Mary Cleave, the trailblazer who made history as the first woman to lead a space mission post the tragic Challenger explosion, passed away at 76, as confirmed by NASA on Monday. A pioneer in space exploration and environmental engineering, Cleave's impactful career included deploying the Magellan Venus probe and overseeing key programs at NASA's Science Mission Directorate. Her legacy is marked by achievements, including being the first woman to fly after the Challenger tragedy, contributing to spacewalks, and advancing scientific exploration. NASA Associate Administrator Bob Cabana expressed sorrow, highlighting Cleave's passion for science and dedication to Earth. Born in Southampton, New York, in 1947, Mary Cleave's contributions earned her accolades, including NASA Space Flight and Exceptional Service medals. She will be dearly missed.